Contents

`IØ162961`

PAGE   PARA-GRAPH

Background. 6

Part 1. The psychosis-addressing-and-solving, real biological explanation of the human condition . . . 9   1-

Part 2. Since this instinct vs intellect explanation is reasonably obvious, why haven't we been taught it at school?. 25   46-

Part 3. How did we humans acquire our instinctive cooperative and loving moral conscience?. 38   79-

Part 4. How does the psychological rehabilitation of the human race that the arrival of understanding of the human condition finally makes possible actually occur?. 47   104-

---

THE Interview **is free to watch, hear, read and print at HumanCondition.com, and you can:**

1. Purchase copies from bookshops, including Amazon, for around $US11 each (in 2020). These are beautiful perfect bound, glossy-covered, colour booklets.

2. Purchase promotional copies at a very low price from the WTM for giving away, minimum order of 20. These are also the beautiful perfect bound, glossy-covered, colour booklets but have a different ISBN to the bookshop version and have stated on their covers 'PROMOTIONAL COPY, NOT FOR RESALE'. The price of these promotional booklets is around $US4 (in 2020) each which *includes* delivery to most places in the world.

3. Print your own copies as either A4 or US letter PDF, or the smaller A5 or Half US letter PDF for easy carry & distribution, but these stapled folded booklets look more like pamphlets than the lovely books above.

See details at www.humancondition.com/the-interview-book.

Background

© 2020 Fedmex Pty Ltd

About Jeremy Griffith (pictured right)

Jeremy is an Australian biologist who has dedicated his life to bringing fully accountable, biological understanding to the dilemma of the human condition; which is the underlying issue in all human life of our species' extraordinary capacity for what has been called 'good' and 'evil'.

Jeremy has published over ten books on the human condition, including the Australasian bestseller *A Species In Denial* (2003), and his definitive treatise, *FREEDOM: The End Of The Human Condition* (2016).

His work has attracted the support of such eminent scientists as the former President of the Canadian Psychiatric Association Professor Harry Prosen, the esteemed American ecologist Professor Stuart Hurlbert, Australia's Templeton Prize-winning biologist Professor Charles Birch, the Former President of the Primate Society of Great Britain Dr David Chivers, Nobel Prize winning physicist Stephen Hawking, as well as other distinguished thinkers such as Sir Laurens van der Post—see www.humancondition.com/#commendations.

Jeremy is the founder and patron of the World Transformation Movement (WTM)—see www.humancondition.com.

Commendations for Griffith's treatise
From Thought Leaders

'[**Prof. Stephen Hawking**] is most interested in your impressive proposal.'
• 'In all of written history there are only 2 or 3 people who've been able to think on this scale about the human condition.' **Prof. Anthony Barnett**, zoologist
• '*FREEDOM* is the book that saves the world...cometh the hour, cometh the man.' **Prof. Harry Prosen**, former Pres. Canadian Psychiatric Assn. • 'I am stunned and honored to have lived to see the coming of "Darwin II".' **Prof. Stuart Hurlbert**, esteemed ecologist • 'Living without this understanding is like living back in the stone age, that's how massive the change it brings is!' **Prof. Karen Riley**, clinical pharmacist • 'Frankly, I am blown away by the ground-breaking significance of this work.' **Prof. Patricia Glazebrook**, philosopher • 'I've no doubt a fascinating television series could be made based upon this.' **Sir David Attenborough** • '*FREEDOM* is the necessary breakthrough in the critical issue of needing to understand ourselves.' **Prof. David J. Chivers**, former Pres. Primate Society of Britain • 'Whack! Wham! I was converted by Griffith's erudite explanation for our behaviour.' **Macushla O'Loan**, *Executive Women's Report* • 'This is indeed impressive.' **Dr Roger Lewin**, preeminent science writer • 'I have recommended Griffith's work for his razor-sharp biological clarifications.' **Prof. Scott Churchill**, psychologist • 'An original and inspiring understanding of us.' **Prof. Charles Birch**, zoologist • 'The insights are fascinating and pertinent and must be disseminated.' **Dr George Schaller**, pre-eminent biologist • 'Very impressive, particularly liked the primatology section.' **Prof. Stephen Oppenheimer**, geneticist, author *Out of Eden* • 'I consider the book to be the work of a prophet.' **Dr Ron Strahan**, former dir. Sydney Taronga Zoo • 'The scholarly value [of Griffith's synthesis] is comparable to several of the most celebrated publications in biology.' **Prof. Walter Hartwig**, anthropologist • 'I believe you're on to getting answers to much that has bewildered humans.' **Dr Ian Player**, famous Sth. Afr. conservationist • 'A superb book, a forward view of a world of humans no longer in naked competition.' **Prof. John Morton**, zoologist • 'This might bring about a paradigm shift in the self-image of humanity.' **Prof. Mihaly Csikszentmihalyi**, psychologist • 'As a therapist this is a simply brilliant explanation.' **Jayson Firmager**, founder of *Holistic Therapist Magazine* • 'The questions you raise stagger me into silence; most admirable.' **Ian Frazier**, author *Great Plains* bestseller • 'The WTM is an island of sanity in a sea of madness.' **Tim Macartney-Snape**, world-leading mountaineer & twice Order of Australia recipient

Commendations From The General Public

'Griffith should be given Nobel prizes for peace, biology, medicine; actually every Nobel prize there is!' ● 'He nailed it, nailed the whole thing, just like the world going from FLAT to ROUND, BOOM the WHOLE WORLD CHANGES, no joke.' ● '*FREEDOM* will be the most influential, world-changing book in history, and time will now be delineated as BG, before Griffith, or AG, after Griffith.' ● 'I'm speechless – this is bigger than natural selection & the theory of relativity!' ● 'I really think this man will become recognized as the best thinker this world's ever seen, and don't we need him right now!' ● 'Griffith has decoded the human species, we FINALLY know what's going on & the suffering stops!' ● 'The world can't deny this for much longer, let the light in, save the human race!' ● 'This is the most exciting moment in my life. *THE Interview* tore my hat off & let my brain fly into the sky!' ● '*THE Interview* should be globally broadcast daily. The healing explanation humans so sorely need.' ● 'In a world that's lost its way there's no greater breakthrough, water to a world dying of thirst.' ● 'Dawn has come at Midnight! A brilliant exposition, we could be on the cusp of regaining Paradise!' ● 'This man has broken the great silence, defeated our denial, got the truth up, woken us from a great trance.' ● 'Beware the 'deaf effect; your mind will initially resist the issue of our corrupted condition and so find it hard to take in or hear what's being said, but if you're patient you'll find the redeeming explanation of our condition pure relief.' ● 'John Lennon pleaded "just give me some truth", well this site finally gives us *all* the truth!' ● '*FREEDOM* is the most profound book since the Bible, now with the redeeming truth about us humans.' ● '*Death by Dogma* is brilliant clarification.' ● 'We were given a computer brain, but no program for it; but Aha, Griffith has found it, made sense of our lives!' ● 'This just goes deeper & deeper in explaining us, like dawn devouring darkness, amazing!' ● 'Agree, this is not another deluded, pseudo idealistic, PC, 'woke', false start to a better world, but the human-condition-resolved real solution.' ● 'Freedom indeed! What we have here is the second coming of innocence who exposes us but sets us free!' ● 'As prophesised, King Arthur has returned to save us (mentioned in par.1036 *Freedom*)' ● 'We all need to go back to school & learn this truthful explanation of life.' ● 'Join in our jubilation, your magic reunites, all men become brothers, all good all bad, be embraced millions! This kiss [of understanding] for the whole world' – From Beethoven's 9th (par.1049 *Freedom*)

THE Interview That Solves The Human Condition And Saves The World!

The transcript of acclaimed British actor and broadcaster Craig Conway's astonishing, world-changing and world-saving 2020 interview with Australian biologist Jeremy Griffith about his book *FREEDOM: The End Of The Human Condition*—which presents the completely redeeming, uplifting and healing understanding of the core mystery and problem about human behaviour of our so-called 'good and evil'-stricken *human condition*—thus ending all the conflict and suffering in human life at its source, and providing the now urgently needed road map for the complete rehabilitation and transformation of our lives and world!

THE Interview was broadcast across the UK in 2020 and is being replayed on radio and TV stations around the world.

This booklet provides the ideal, very short, introductory summary of Jeremy Griffith's book *FREEDOM: The End Of The Human Condition*.

The film of this world-saving interview can be watched at HumanCondition.com

OR

Scan code to view

THE Interview That Solves The Human Condition And Saves The World!
by Jeremy Griffith and Craig Conway

Published in 2020, by WTM Publishing and Communications Pty Ltd
(ACN 103 136 778) (www.wtmpublishing.com).

All enquiries to:

WORLD TRANSFORMATION MOVEMENT®
Email: info@worldtransformation.com
Website: www.humancondition.com or www.worldtransformation.com

The World Transformation Movement (WTM) is a global not-for-profit movement
represented by WTM charities and centres around the world.

ISBN 978-1-74129-056-1
CIP – Biology, Philosophy, Psychology, Health

Filming of Jeremy Griffith and the editing and production of the film by
James Press.

© 2020 Fedmex Pty Ltd

About *FREEDOM: The End Of The Human Condition*

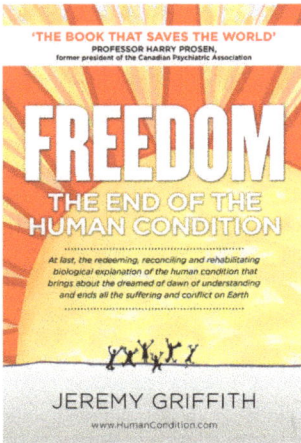

Jeremy's book *FREEDOM*, which this interview is about, presents the reconciling and psychologically healing biological explanation of our 'good and evil' stricken *human condition*. In doing so it unravels the core mystery behind human behaviour, thus ending suffering and conflict at its source—and providing the now urgently needed road map for the complete transformation of our lives and world.

FREEDOM was launched at the Royal Geographical Society in London in 2016, with the keynote address by Sir Bob Geldof. It is freely available at www.humancondition.com, or you can purchase hard copies at bookstores, including Amazon.

About Craig Conway

Craig is an internationally acclaimed English actor, writer, producer and director who has worked in theatre, TV and film for three decades. On stage Craig has created projects for theatre companies including The National Theatre, The Royal Shakespeare Company, The Northern Stage Ensemble (Founder member) and the Contact Theatre Manchester. On screen, he has starred in films such as *Final Score* with Pierce Brosnan, *The Current War* with Benedict Cumberbatch and *The Courier* with Gary Oldman. His TV credits include *Vera*, *Wire In the Blood*, *George Gently* and *Our Friends In The North*. Craig is a member of the Writers' Guild of Great Britain.

With the onset of the pandemic in 2020, Craig decided to try to reach people through radio and conducted a series of widely broadcast interviews with people he felt were **'seeking to make a difference in the world'**, which included this interview with Jeremy.

Craig first learnt of Jeremy's explanation of the human condition in early 2019 and became so impressed by it he started a WTM Centre in north east England to promote it; see www.WTMNorthEastEngland.com.

Part 1

The psychosis-addressing-and-solving, real biological explanation of the human condition

[1] **Craig Conway**: Hello to everyone listening. My name is Craig Conway. Now, whilst I've been an actor by profession, very recently I've been introduced to doing radio where I talk to people from all over the world. Well, today I have a very, very special guest on the line from Australia.

[2] The turmoil and trauma of this pandemic has only amplified the now dire need in the world for a deeper, lasting solution to all the chaos and suffering in human life. And this deeper enduring solution is *actually* what this biologist I'm about to interview is going to provide us with. He is going to do it by explaining and solving the underlying cause of all the suffering, which is our 'good and evil' stricken so-called *human condition*.

[3] So I don't care what you're doing, you need to stop and listen to this interview. In fact, I don't care what you do for the rest of your life, if you can you just need to listen to this!

[4] The interview will be in four parts, each averaging 15 minutes, which is not a lot when you consider that we're going to be explaining the whole human condition!

[5] So it's a great privilege to introduce Australian biologist Jeremy Griffith. He's the author of a book titled *FREEDOM: The End Of The Human Condition*, and this is my copy, which I've had with me now for quite a long time and I take it *everywhere* with me, and there are now millions of people all over the globe studying, reading and researching through this book that Jeremy has brought to us. [Craig learnt of Jeremy's explanation of the human condition in early 2019,

and became so impressed by it he started a WTM Centre in north east England to promote it; see www.WTMNorthEastEngland.com.]

[6] So, I'm here to tell everyone that this book has not only blown me away, it has also impressed Professor Harry Prosen, who is a former president of the Canadian Psychiatric Association—so he's one of the world's leading psychiatrists—and he said, and I quote, '**I have no doubt Jeremy Griffith's biological explanation of the human condition is the holy grail of insight we have sought for the psychological rehabilitation of the human race. This is the book we have been waiting for, it is the book that saves the world.**' End quote.

[7] Now, I think everyone listening would agree that '**the psychological rehabilitation of the human race**' is exactly what this world needs! So buckle into your seats, this is going to be the most interesting—and exciting—talk you have ever heard.

[8] So Jeremy, thank you for talking with us. Tell us, how does your work bring about '**the psychological rehabilitation of the human race**' and end all the suffering and strife, and, as Professor Prosen said, '**save the world**'?

[9] **Jeremy Griffith**: Thank you very much for having me on your program Craig. Finding understanding of our psychologically troubled human condition has actually been what the efforts of every human who has ever lived has been dedicated to achieving and has contributed to finding. As Professor Prosen said, finding understanding of the human condition has been '**the holy grail**' of the whole human journey of conscious thought and enquiry.

[10] We humans have absolutely lived in hope, faith and trust that one day, somewhere, some place, all the efforts of everyone—but of scientists in particular—would finally produce the completely redeeming, uplifting and healing understanding of us humans. I know it must seem outrageous to claim that this goal of goals has finally been achieved, but it has. In fact, the human condition is such a difficult subject for us humans to confront and deal with that I couldn't be talking about it so openly and freely if it *hadn't* been solved.

[11] **Craig**: Okay then Jeremy, solve the human condition for us, we're all ears!

[12] **Jeremy**: Firstly, I'm a biologist, and that's important because I think everyone will agree that what we need is a non-abstract, non-mystical, completely rational and thus understandable, scientific, biological explanation of us humans.

[13] So how are we to explain and understand *the human condition*, understand why we humans are the way we are, so brutally competitive, selfish and aggressive that human life has become all but unbearable. In fact, how are we to make *so much* sense of our divisive behaviour that the underlying cause of it is so completely explained and understood that, as Professor Prosen said, the whole of the human race is psychologically rehabilitated and everyone's life is transformed?

[14] **Craig**: Yes, that's what we want; the human condition finally explained, fixed up and healed *forever*!

[15] **Jeremy**: Exactly Craig. So, to start at the beginning, I know everyone listening is living with the belief—well it's what we were all taught at school and are told in every documentary—that humans' competitive, selfish and aggressive behaviour is due to us having savage, must-reproduce-our-genes instincts like other animals have. Certainly while left-wing thinkers do claim we have some selfless, cooperative instincts, they also say we have this selfish, competitive 'animal' side, which Karl Marx limited to such basic needs as sex, food, shelter and clothing.

[16]I mean, our conversations are saturated with this belief, with comments like: 'We are programmed by our genes to try to dominate others and be a winner in the battle of life'; and 'Our preoccupation with sexual conquest is due to our primal instinct to sow our seeds'; and 'Men behave abominably because their bodies are flooded with must-reproduce-their-genes-promoting testosterone'; and 'We want a big house because we are innately territorial'; and 'Fighting and war is just our deeply-rooted combative animal nature expressing itself'.

[17]**Craig**: Yes, that's exactly what I've understood is the reason for our competitive and aggressive nature—that we have brutally competitive, survival-of-the-fittest instincts, which we are always having to try to restrain or civilise or try to control as best we can; I mean that's what I was taught in school.

[18]**Jeremy**: Yes, that's what we were taught, but let's think about this—and what I'm going to say now is very important, so I hope everyone's listening closely.

[19]Surely this idea that we have savage competitive and aggressive, must-reproduce-our-genes instincts *cannot* be the real reason for our species' competitive and aggressive behaviour because, after all, words used to describe our human behaviour such as egocentric, arrogant, inspired, depressed, deluded, pessimistic, optimistic, artificial, hateful, cynical, mean, sadistic, immoral, brilliant, guilt-ridden, evil, psychotic, neurotic and alienated, all recognise the involvement of OUR species' fully conscious thinking mind. They demonstrate that there is a *psychological* dimension to our behaviour; that we don't suffer from a genetic-opportunism-driven 'animal condition', but a conscious-mind-based, *psychologically* troubled HUMAN CONDITION.

Our ape ancestors were *not* savage, barbaric brutes as they have for so long been portrayed, but rather they were innocent, loving nurturers as depicted overleaf by the paleoartist Jay H. Matternes in *Science* magazine.

[20] What's more, we humans have cooperative, selfless and loving *moral* instincts, the voice or expression of which we call our con-science—which is the complete *opposite* of competitive, selfish and aggressive instincts. As Charles Darwin said, **'The moral sense…affords the best and highest distinction between man and the lower animals'** (*The Descent of Man*, 1871, ch.4). Of course, to have acquired these cooperative, selfless and loving moral instincts our distant ape ancestors must have *lived* cooperatively, selflessly and lovingly, otherwise how else could we have acquired them? Our ape ancestors can't have been brutal, club-wielding, competitive and aggressive savages as we have been taught, rather they must have lived in a Garden of Eden-like state of cooperative, selfless and loving innocent gentleness—which, as I'd like to explain to you later in this interview Craig, is a state that the

bonobo species of ape is currently living in, and which anthropological findings now evidence we did once live in. For instance, anthropologists like C. Owen Lovejoy are reporting that **'our species-defining cooperative mutualism can now be seen to extend well beyond the deepest Pliocene** [which is well beyond 5.3 million years ago]' ('Reexamining Human Origins in Light of *Ardipithecus ramidus*', *Science*, 2009, Vol.326, No.5949).

Group of bonobos

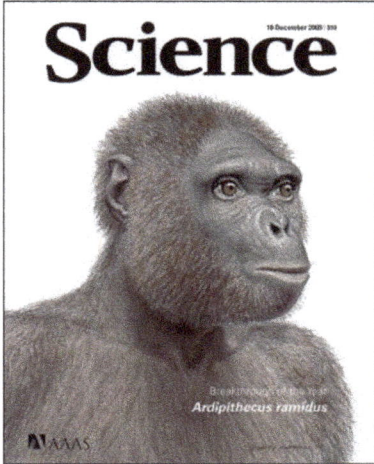

'Breakthrough of the Year':
cover of the December 2009
issue of *Science* magazine

Matternes's reconstruction of the
4.4 mya *Ardipithecus ramidus*
in its natural habitat

[21] So saying our competitive and aggressive behaviour comes from savage competitive and aggressive instincts in us is simply not true—as I'd like to come back to shortly, it's just a convenient excuse we have used while we waited for the psychosis-acknowledging-and-solving, *real* explanation of our present competitive and aggressive human condition!

[22] **Craig**: Wow, so that's a pretty big statement Jeremy, I mean it's a pretty important point you're making here. You're saying that our competitive and aggressive behaviour is *not* due to must-reproduce-our-genes instincts like other animals, but is due to a conscious-mind-based, psychologically troubled condition, yes?

[23] **Jeremy**: Yes, our egocentric and arrogant and mean and vindictive and even sadistic behaviour has nothing to do with wanting to reproduce our genes. *That was absurd*. And it is actually really good news that our behaviour is due to a conscious-mind-based psychologically troubled condition because psychoses can be healed with

understanding. If our competitive and aggressive behaviour was due to us having savage instincts then we would be stuck with that born-with, hard-wired, innate behaviour. It would mean we could only ever hope to restrain and control those supposedly brutal instincts. But since our species' divisive behaviour is due to a psychosis, that divisive behaviour *can* be cured with healing understanding. So that *is* very good news. In fact, incredibly exciting news, because <u>with understanding we can finally end our *psychologically* troubled human condition. It's the understanding of ourselves that we needed to heal the pain in our brains and become sound and sane again.</u>

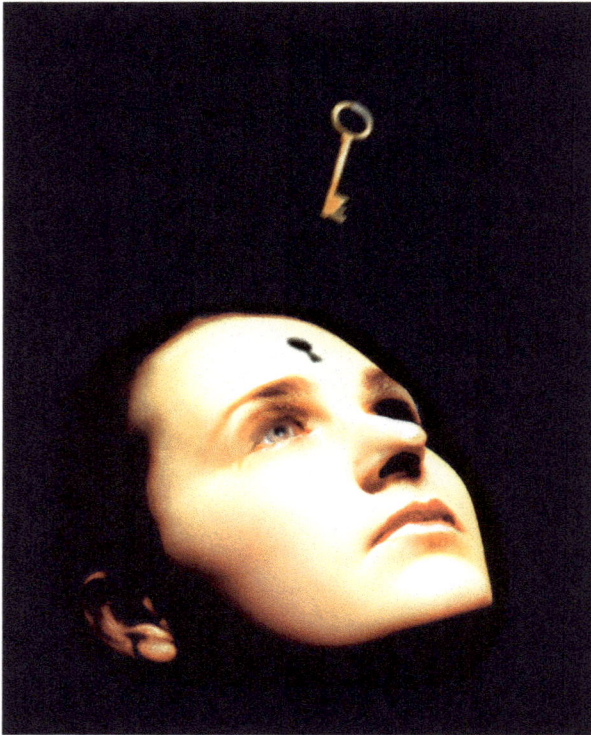

[24] As I said, the 'savage instincts' explanation was just a convenient excuse while we searched for the psychosis-addressing-and-solving *real* explanation of our divisive behaviour, <u>which is the explanation I would now like to present</u>.

[25] **Craig**: Okay, so what you're saying here, Jeremy, is that we don't need the convenient excuse anymore that we have some kind of savage animal instincts because we have the real explanation of our conscious-mind-based psychologically troubled human condition!

[26] **Jeremy**: Yes, and this key, all-important, psychosis-addressing-and-solving explanation is actually very obvious.

[27] If we think about it, if an animal was to become fully conscious, like we humans became, then that animal's new self-managing, understanding-based conscious mind would surely have to challenge its pre-existing instinctive orientations to the world, wouldn't it? A battle would *have to* break out between the emerging conscious mind that operates from a basis of understanding cause and effect and the non-understanding instincts that have always controlled and dictated how that animal behaves.

[28] **Craig**: Yes, that makes sense Jeremy, so what happened though when this animal became conscious and its whole life turned into a psychologically distressed mess?

[29] **Jeremy**: Well, the easiest way to see what happened is to imagine the predicament faced by an animal whose life had always been controlled by its instincts suddenly developing a conscious mind, because if we do that we will very quickly see how that animal would develop a psychologically troubled competitive and aggressive condition like we suffer from.

[30] So let's imagine a stork: we'll call him Adam. Each Summer, Adam instinctually migrates North with the other storks around the

coast of Africa to Europe to breed, as some varieties of storks do. Since he has no conscious mind Adam Stork doesn't think about or question his behaviour, he just follows what his instincts tell him to do.

ADAM

Drawing by Jeremy Griffith © 1991-2016 Fedmex Pty Ltd

[31] But what if we give Adam a large brain capable of conscious thought? He will start to think for himself, but many of his new ideas will not be consistent with his instincts. For instance, while migrating North with the other storks Adam notices an island full of apple trees. He then makes a conscious decision to divert from his migratory path and explore the island. It's his first grand experiment in self-management.

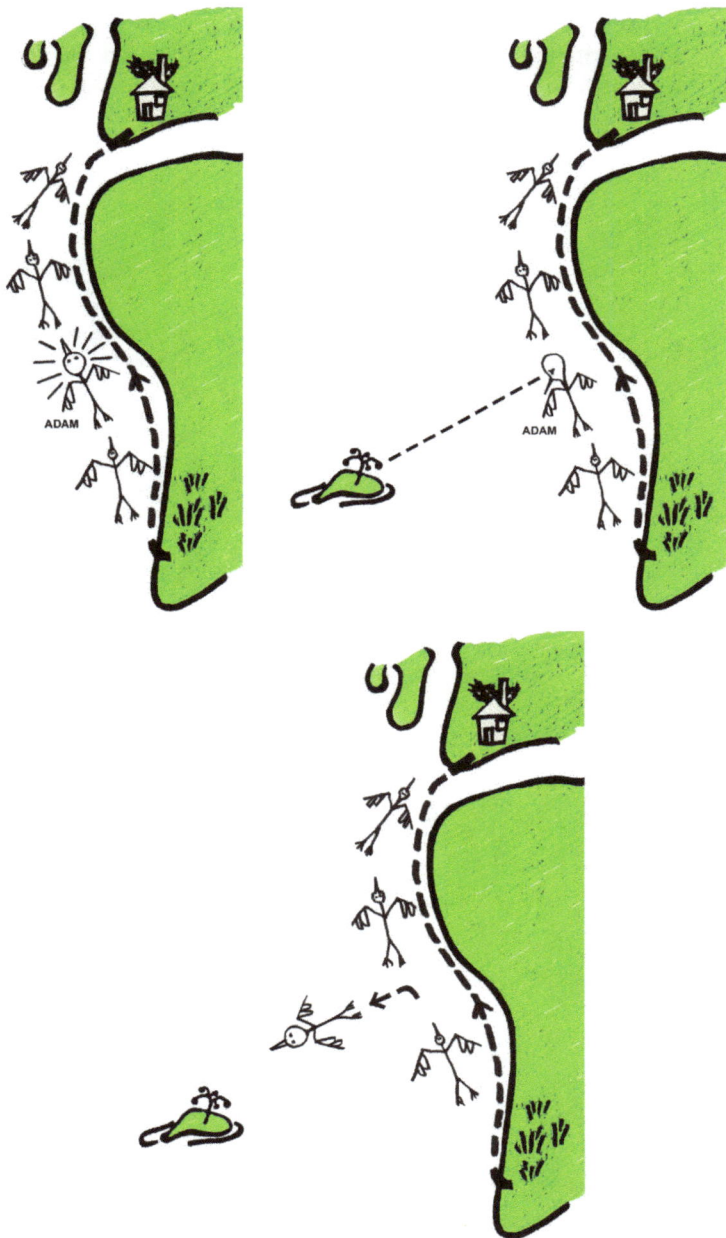

[32] But when Adam's instincts realise he has strayed off course they are going to criticise his deprogrammed behaviour and dogmatically try to pull him back on his instinctive flight path, aren't they! In effect, they are going to condemn him as being bad.

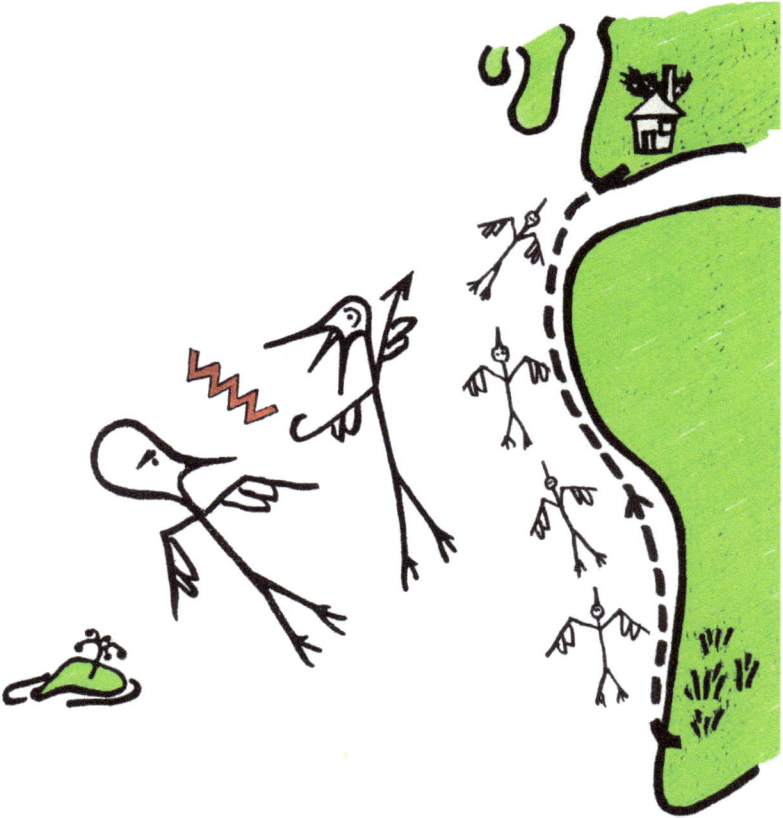

[33] Imagine the turmoil Adam will experience: he can't go back to simply following his instincts. His instinctive orientations to the migratory flight path were acquired over thousands of generations of natural selection but those orientations are not understandings, and since his conscious mind requires understanding, which it can only

get through experimentation, inevitably a war will break out with his instincts.

[34] Ideally at this point Adam's conscious mind would sit down and explain to his instincts *why* he's defying them. He would explain that the gene-based, natural selection process only gives species instinctive *orientations* to the world, whereas his nerve-based, conscious mind, which is able to make sense of cause and effect, needs *understanding* of the world to operate.

[35] But Adam doesn't have this self-understanding. He's only just begun his search for knowledge. In fact, he's not even aware of what the problem actually is. He's simply started to feel that he's bad, even evil.

[36] **Craig**: Okay, so what you're saying is a war has broken out between his conscious mind and his instincts, which he can't explain, and it's left him feeling bad or that he *is* bad in some way, or even evil. So what happened then?

[37] **Jeremy**: Well, tragically, while searching for understanding, we can see that three things are *unavoidably* going to happen. Adam is going to defensively retaliate against the implied criticism from his instincts; he is going to desperately seek out any reinforcement he can find to relieve himself of the negative feelings; and he is going to try to deny the criticism and block it out of his mind. He has become angry, egocentric and alienated—which is the psychologically upset state we call the human condition, because it was us humans who developed a conscious mind and became psychologically upset. (And 'upset' is the right word for our condition because while we are not 'evil' or 'bad', we are definitely psychologically upset from having to participate in humanity's heroic search for knowledge. 'Corrupted' and 'fallen' have been used to describe our condition, but they have negative connotations that we can now appreciate are undeserved, so 'upset' is a better word.)

[38] So Adam's intellect or 'ego' (ego being just another word for the intellect since the *Concise Oxford Dictionary* defines **'ego'** as **'the conscious thinking self'** (5th edn, 1964)) became 'centred' or focused on the need to justify itself—Adam became ego-centric, selfishly preoccupied aggressively competing for opportunities to prove he is good and not bad, to validate his worth, to get a 'win'; to essentially eke out any positive reinforcement that would bring him some relief from his criticising instincts. He unavoidably became self-preoccupied or selfish, and aggressive and competitive.

Drawing by Jeremy Griffith © 1991 Fedmex Pty Ltd

[39] So our selfish, competitive and aggressive behaviour is not due to savage instincts but to a psychologically upset state or condition.

[40] Basically suffering psychological upset was the price we conscious humans had to pay for our heroic search for understanding. In the words from the song *The Impossible Dream* from the musical the *Man of La Mancha*, we had to be prepared to **'march into hell for a heavenly cause'** (lyrics by Joe Darion, 1965). We had to lose ourselves to find ourselves; we had to suffer becoming angry, egocentric and alienated until we found sufficient knowledge to explain ourselves.

[41] **Craig**: Wow Jeremy, I mean this is just fascinating. So Adam Stork—we humans—developed a conscious mind and unavoidably started warring with our instincts, an upsetting war which could only end when we could explain and understand why we had to defy our instincts, which is the understanding that you have just supplied, yes?

[42] **Jeremy**: Exactly, remember Adam Stork became defensively angry, egocentric and alienated because he couldn't explain why he was defying his instincts, so now that we *can* explain why, those defensive behaviours are no longer needed and can end!

[43] That's basically all there is to explain, that is the biological explanation of the human condition that so explains us that, as Professor Prosen said, it brings about **'the psychological rehabilitation of the human race'**!

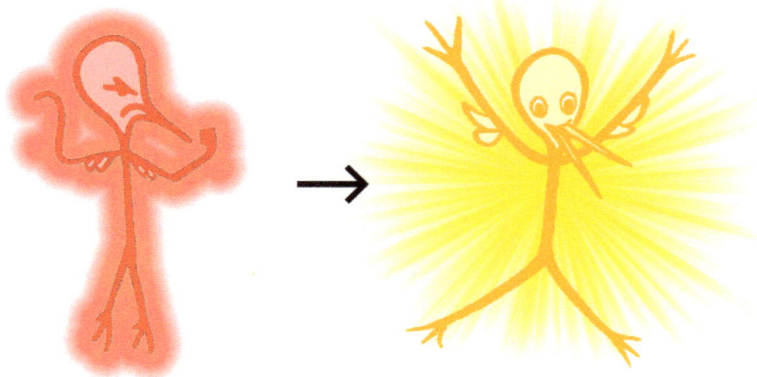

Drawing by Jeremy Griffith © 2016 Fedmex Pty Ltd

[44] **Craig**: This is such a simple story but so far-reaching in its ramifications—I mean it is world-changing is what it is, because it truly enables **'the psychological rehabilitation of the human race'**! I mean that is just wonderful.

[45] Okay, I'm speaking with Australian biologist Jeremy Griffith.

Part 2

Since this instinct vs intellect explanation is reasonably obvious, why haven't we been taught it at school?

[46] **Craig**: Hello, I'm Craig Conway and I'm talking with Australian biologist Jeremy Griffith about how we can end all the turmoil and trauma in the world through explaining and understanding the human condition—which is the instinct vs intellect explanation Jeremy has just given us in Part 1 of this interview.

[47] I do have some questions Jeremy. Firstly, while it seems an obvious explanation that when we became conscious a psychologically upsetting battle must have broken out with our dictatorial instincts, but if it is so obvious, why weren't we taught this at school? And secondly, how were our bonobo-like ape ancestors able to become cooperative and loving, which, as you said, must be the origin of our instinctive moral conscience that Darwin said distinguishes us from other animals? And my third question is, how does **'the psychological rehabilitation of the human race'** that Professor Prosen describes actually take place; I mean do we all have to go into therapy or something?

[48] Okay, by the way, I want to mention also that since there's quite a few new—and very interesting—concepts to think about, Jeremy has said that he'll make both the video and the transcript of this interview available at the top of the website that promotes this understanding of the human condition, which is the World Transformation Movement's website at HumanCondition.com. Now this interview will be there as a video, and the transcript as a little free book, so you can re-listen to, or read the interview again there, because, as I said, with this big subject of the human condition, there is quite a bit to take in and think about.

[49] **Jeremy**: Okay, they are very good questions Craig.

[50] So, to begin with your first question, which is that if this instinct vs intellect explanation is so obvious why haven't we been taught it at school?

⁵¹ The answer is that while it has been recognised—even from ancient times—that the emergence of our conscious mind somehow caused us to 'fall from grace', or however you want to describe the corruption of our original innocent cooperative, selfless and loving instinctive state, it wasn't until science revealed the difference between the gene-based and nerve-based learning systems—which is that genes can orientate but nerves need to understand—that we were finally in a position to explain the good reason for our angry, egocentric and alienated human condition.

⁵² The Biblical story of Adam and Eve in the Garden of Eden that Moses wrote so long ago in about 1,500 BC actually perfectly describes the psychologically upsetting battle that emerged between our instincts and conscious intellect. It says Adam and Eve/we humans took the **'fruit'** (Genesis 3:3) **'from the tree of knowledge'** (Gen. 2:9, 17) and were **'disobedient'** (the term widely used in descriptions of Gen. 3). In other words, we developed a conscious mind and free will. But in that *pre-scientific* story it says Adam and Eve then became **'evil'** (Gen. 3:22) perpetrators of **'sin'** (Gen. 4:7) because they became angry, egocentric and alienated, and as a result Moses said they were **'banished...from the Garden of Eden'** (Gen. 3:23) state of cooperative and loving innocence.

Adam and Eve,
by Lucas Cranach the Elder (1526)

Adam and Eve cast out of Paradise, from *Old Testament Stories* pub. Society for Promoting Christian Knowledge, London (c.1880)

[53] You see, not knowing how naturally selected instincts differ from cause and effect-operating consciousness, this story of Adam and Eve becoming conscious could only conclude that the angry, egocentric and alienated condition that emerged when we became conscious was a bad, evil, sinful state, but this *scientific* presentation says, 'No, no, that pre-scientific story got it wrong'. Adam and Eve are actually not just good but the heroes of the whole story of life on Earth—because surely the conscious mind is nature's greatest invention and to be given the task of searching for understanding while the whole world's condemning you was the hardest and toughest of tasks—because that condemnation *was* universal. All the other innocent storks are condemning the search for knowledge, and since all of nature—the rain, the clouds, the trees, and other animals—are all associated with our original instinctive self that was condemning us, the whole world, in effect, ganged up on Adam and Eve, i.e. us humans—and yet all the time we were good and not bad but we couldn't explain why, but now at last through the benefit of science, we can.

[54] **Craig**: Yeah, I hadn't realised that, but it is true, I mean Adam and Eve taking the fruit from the tree of knowledge is a metaphor for becoming conscious, and then they were thrown out of the Garden of Eden of original innocence because it appeared that they were bad for doing so, but now thankfully we can explain that they, we humans, weren't bad at all; in fact, we are the heroes of the story of life on Earth!

[55] **Jeremy**: That's right, we can now explain and understand that we conscious humans are immense heroes, and not villains after all. How relieving is that!

Richard Heinberg (1950–)

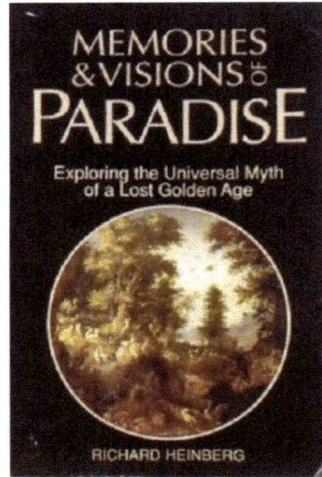

[56] And with regard to recognition of the upsetting conflict between our moral instincts and our conscious intellect, the Biblical story of Adam and Eve is far from the only recognition of it from ancient times. Indeed, as the researcher Richard Heinberg summarised in his 1990 book *Memories & Visions of Paradise* (underlinings are Jeremy's emphasis), **'Every religion begins with the recognition that <u>human consciousness has been separated from the divine Source, that a former sense of oneness… has been lost</u>…everywhere in religion and myth there is an acknowledgment that we have departed from an original…innocence and can return to it only through the resolution of some profound inner discord…the cause of the Fall is described variously as disobedience, as the eating of a forbidden fruit** [from the tree of knowledge]**, and as spiritual amnesia** [forgetting, blocking out, denial, alienation, which is our psychosis]**'** (pp.81-82 of 282). So all our religions and most of our mythologies have recognised the basic conflict within us—that the emergence of **'consciousness'** caused our **'Fall'** from **'innocence'**.

Hesiod, from the Monnus mosaic, c.200

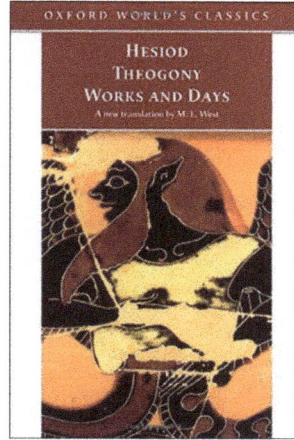

[57] Way back in about 800 BC, the Greek poet Hesiod wrote of our species' pre-conscious time of living cooperatively and lovingly in his epic poem *Works and Days*: '**When gods alike and mortals rose to birth / A golden race the immortals formed on earth...Like gods they lived, with calm untroubled mind / Free from the toils and anguish of our kind / Nor e'er decrepit age misshaped their frame...Strangers to ill, their lives in feasts flowed by...Dying they sank in sleep, nor seemed to die / Theirs was each good; the life-sustaining soil / Yielded its copious fruits, unbribed by toil / They with abundant goods 'midst quiet lands / All willing shared the gathering of their hands'** (*The Remains of Hesiod the Ascræan,* tr. C.A. Elton, pp.17-18). So yes, they didn't have a troubled conscious mind, and they lived a sharing, gentle life.

[58] **Craig**: Yes, I've heard of the idea of a 'golden race', but I didn't actually know where it came from. So what you're saying then, Jeremy, is our distant ancestors had a **'calm untroubled mind'**—no human condition yet!

[59] **Jeremy**: Yes, that's right, and in 360 BC Hesiod's Greek compatriot Plato gave this very similar description of our species' pre-conscious time in innocence. He wrote: **'there was a time when…we beheld the beatific vision and were initiated into a mystery which may be truly called most blessed, celebrated by us in <u>our state of innocence, before we had any experience of evils to come, when we were admitted to the sight of apparitions innocent and simple and calm and happy,</u> which we beheld shining in pure light, pure ourselves and not yet enshrined in that living tomb which we carry about, now'** (*Phaedrus*; tr. B. Jowett, 1871, 250).

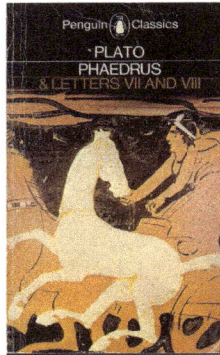

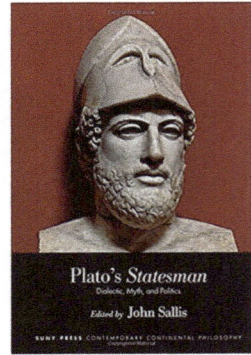

Plato, c.428–348 BC; *Phaedrus*; *Statesman*

[60] Plato also gave this other description of the innocent 'Golden Age' in our species' pre-conscious past, writing of a time when we lived a **'blessed and spontaneous life…**[where] **neither was there any violence, or devouring of one another, or war or quarrel among them…In those days God himself was their shepherd, and ruled over them** [in other words, our original instinctive self was orientated to living in an ideal cooperative, loving way]**…Under him there were no forms of government or separate possession of women and children; for all men rose again from the earth, having <u>no memory of the past</u>** [in other words, we lived in a pre-conscious state]**. And…the earth gave them fruits in abundance, which grew on trees and shrubs unbidden, and were not planted by the hand of man. And they dwelt naked, and mostly in the open air, for the temperature of their seasons**

was mild; and they had no beds, but lay on soft couches of grass, which grew plentifully out of the earth' (*The Statesman*, c.350 BC; tr. B. Jowett, 1871, 271-272).

[61] The thing is, Hesiod and Plato, like Moses, were living at a time when science still had to be developed, so they weren't able to provide the redeeming, instincts-can-orientate-but-only-nerves-can-understand, good reason WHY we departed from **'innocence'** and seemingly became 'evil', bad people.

[62] **Craig**: Yes, because there was no science back then.

[63] **Jeremy**: Precisely, it's only in the last 150 years or so that science has given us A: the ability to know that the gene-based natural selection process gives species orientations to the world; and B: the knowledge of our nerves and how they are able to remember events, which, much developed, has led to our mind being able to sufficiently understand the relationship between cause and effect to become conscious of, or aware of, or intelligent about those relationships. So that's only happened in the last 150 years, but since, as explained in pars 705-707 of *FREEDOM*, the fossil record of our ancestors suggests that our large association-cortexed, thinking, fully conscious brain appeared some 2 million years ago, that means for almost all of the 2 million years we have been conscious we have had no ability to explain and understand why we corrupted our original innocent instinctive self or soul. And without that redeeming explanation the only way we could cope with the astronomical guilt of having destroyed 'Eden', has been to deny we ever lived in a cooperative and loving innocent state—and that's exactly where the excuse that we have savage competitive and aggressive instincts like other animals came to our rescue.

[64] And, false as it is, it's been an absolutely brilliant excuse because instead of our instincts being all-loving and thus unbearably condemning of our present non-loving state, they are made out to be vicious and brutal must-reproduce-your-genes instincts like other animals have; *and*, instead of our conscious mind being the instinct-defying cause of our corruption, it was made out to be the blameless mediating 'hero' that had to step in and try to control those supposed vicious instincts

<u>within us!</u> And those who dared to admit the truth of our cooperative and loving past, like Hesiod and Plato, were dismissed as deluded romantics, and the whole idea of an innocent, Edenic past was said to be nothing more than a nostalgia for the security and maternal warmth of infancy, that it was **'never an historical state'** as the Jungian psychologist Erich Neumann said in his book *The Origins and History of Consciousness* (1949, p.15 of 493).

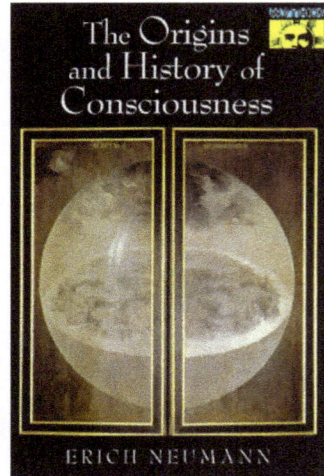

Erich Neumann (1905–1960)

[65] **Craig**: Yes, and we couldn't face the truth that we had turned utopia into dystopia, into a dreadful place of conflict and suffering, yes?

[66] **Jeremy**: Yes that's right Craig, and I should point out that while most contemporary thinkers have clung to the savage instincts excuse for our divisive behaviour, there have been some who, like the ancient thinkers, truthfully recognised the basic instinct vs intellect elements involved in producing the human condition. Eugène Marais, Paul MacLean and Arthur Koestler are a few who come to mind.

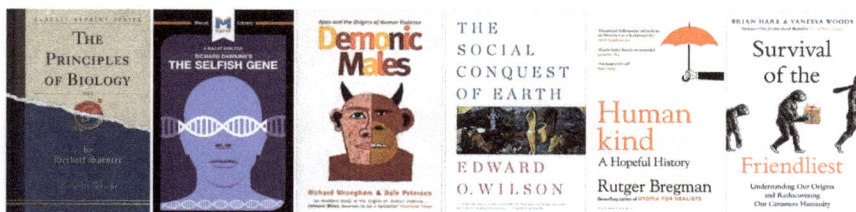

Some of the many science books accounting for human
behaviour using the false 'savage instincts' excuse.

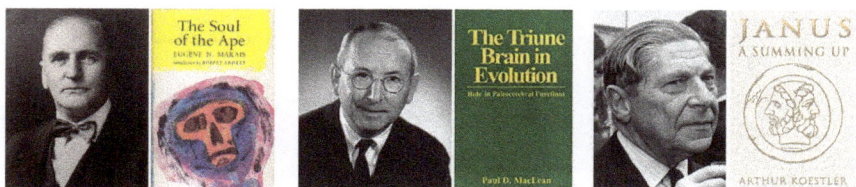

Eugène Marais (1871–1936), Paul MacLean (1913–2007)
and Arthur Koestler (1905–1983)

[67]But while all contemporary thinkers have had the benefit of
science having revealed the difference between the gene-based and
nerve-based learning systems, and have therefore had the means to
truthfully explain the human condition, those who did recognise
the basic instinct vs intellect elements didn't take their thinking far
enough to actually explain the human condition. And those who have
been attached to the false savage instincts excuse—which is the great
majority of scientists—obviously haven't been thinking truthfully, so
they couldn't hope to explain the human condition. Which is all why
it has taken the truthful thinking of the pre-eminent South African
philosopher Sir Laurens van der Post, and following him, myself, to
finally present the complete, true explanation of the human condition.

[68]**Craig**: Okay. And Jeremy, I assume that people can read about
the contemporary thinkers who recognised the instinct vs intellect
elements involved in the human condition, and those who clung to
the savage instincts excuse, on the World Transformation Movement's
website at HumanCondition.com?

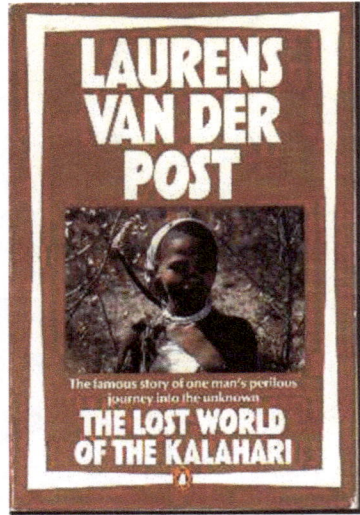

The most famous book by Sir Laurens van der Post (1906–1996) (left) is *The Lost World of the Kalahari*, which is about the relatively innocent Bushmen or San people of the Kalahari Desert. Its title can be interpreted as 'the lost world of our soul'.

[69] **Jeremy**: Yes, they can, in particular in the fourth video at the top of the homepage.

[70] Also, how Sir Laurens van der Post and I managed to address and solve the human condition is described in my 2020 book *How Laurens van der Post Saved The World*, which is also freely available on that website. Basically, what is explained in that book is that since everyone is naturally variously psychologically upset from their different encounters with humanity's battle to find knowledge, there was always going to be a few who were fortunate enough in their infancy and childhood to escape encountering the angry, egocentric and alienated effects of that upsetting battle, and it is these few who could look into the human condition without being overly confronted by it—and Sir Laurens and I were two of these extremely fortunate denial-free thinking people, which is how we were able to find understanding of the human condition.

[71] I should also say that while there is growing support for this now absolutely desperately needed understanding of the human condition, mainstream science is yet to recognise and support it—but that's what happens with paradigm-shifting, breakthroughs in science. When the physicist Max Planck said **'Science progresses funeral by funeral'** (Marilyn Ferguson's reference to a comment by Planck in his *Scientific Autobiography*, 1948; *New Age* mag. Aug. 1982; see www.wtmsources.com/174) he was recognising how attached each generation of scientists becomes to the way of thinking they grew up with, and therefore how slow science as a whole is to move to a new paradigm of understanding. And the playwright George Bernard Shaw also warned of how difficult it is to introduce a new paradigm of thinking—especially one that dares to confront the historically unbearably confronting and off-limits subject of the human condition—when he said that **'All great truths begin as blasphemies'** (*Annajanska*, 1919). So yes, confronting the human condition when everyone has been living in fearful denial of it, even though it has finally been explained and made safe to confront, represents the biggest of all **'blasphemies'**.

[72] **Craig**: Yeah, I understand completely. I mean from my limited experience, Jeremy, I know how difficult it is to get people to change their way of thinking. But your basic point is that science's discovery of the way genes and nerves work has finally made it possible for the liberation of humanity from the horror of the human condition.

[73] **Jeremy**: Yes, that is the essential truth: science is the liberator, the so-called 'messiah' or 'redeemer' of humankind, as we always hoped it would be!

[74] **Craig**: And all this is in the nick of time Jeremy, because I actually don't think the world can cope with any more upset behaviour from us humans—but of course we still need the scientific community to get on board and support this understanding.

[75] **Jeremy**: Absolutely. While we've had to live in denial of our corrupted, psychologically upset condition while we couldn't explain it, the truth is that on a graph showing the ever-increasing levels of upset in humans, those levels, especially of psychosis and alienation, have been increasing so rapidly lately that the line tracking their rise is near vertical with the amount of upset virtually doubling now in each new generation! Freedom Essays 55 & 59 on our World Transformation Movement's website describes this terrifying end play threat of terminal levels of psychosis. Basically, we had virtually lost the race between self-destruction and self-understanding.

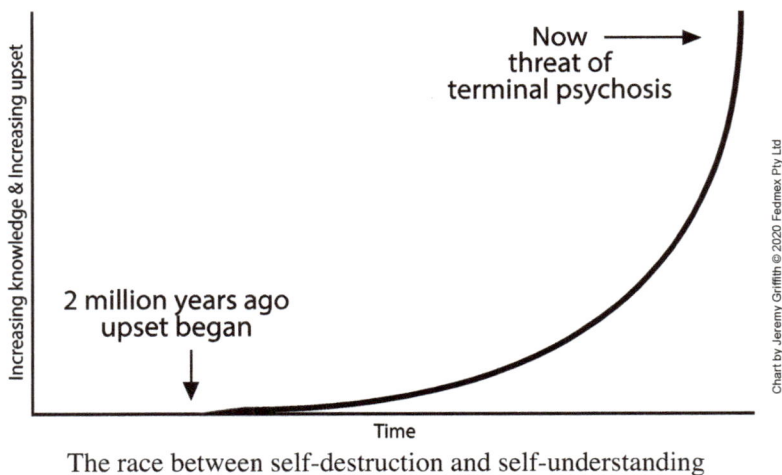

The race between self-destruction and self-understanding

[76] Indeed, early last century the author Antoine de Saint-Exupéry wrote that **'We are living through deeply anxious days, and if we are to relieve our anxiety we must diagnose its cause…What is the meaning of man? To this question no answer is being offered, and I have the feeling that we are moving toward the darkest era our world has ever known'** (*A Sense of Life*, pub. 1965, pp.127, 219 of 231), and the **'deeply anxious days'** have very greatly

increased in the century since then, so we are now very much in the midst of **'the darkest era our world has ever known'**—so this world-saving, **'reliev**[ing]**'**-of-the-**'cause'**, understanding of **'the meaning of man'**, has definitely only arrived in the nick of time—which means the scientific community definitely, definitely needs to get its act together and support this breakthrough!

Antoine de Saint-Exupéry (1900–1944),
depicted on France's 50 franc note

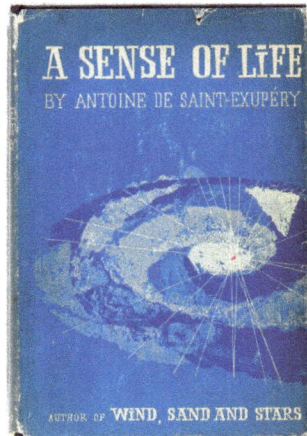

[77] So now I'll address your other questions Craig, but I should mention that the first of the videos on the World Transformation Movement's website warns of the difficulty not only scientists have had, but almost everyone has had trying to confront and think about the historically unbearably confronting subject of our corrupted condition which has finally been explained and made safe to confront and admit.

[78] **Craig**: I'm Craig Conway and I'm speaking with Australian biologist Jeremy Griffith.

Part 3

How did we humans acquire our instinctive cooperative and loving moral conscience?

[79] **Craig**: We're continuing our important conversation with Australian biologist Jeremy Griffith who's been explaining the human condition and how understanding it can end all the trauma and suffering in the world—and boy, don't we need that!

[80] So in Part 1 Jeremy explained that we conscious humans became angry, egocentric and alienated because we couldn't explain why we had to defy our instincts, but now that we can explain this and understand ourselves those defensive ways of coping are no longer needed and the human race is psychologically rehabilitated.

[81] In Part 2 Jeremy explained that the reason we had to use the dishonest savage instincts excuse is because science had to find understanding of the way genes and nerves work, that genes can orientate a species but nerves need to understand cause and effect.

[82] So Jeremy, my second question for this, Part 3, of our interview is how were our bonobo-like ape ancestors able to become cooperative and loving, which, as you said, must be the origin of our instinctive moral conscience that Darwin said distinguishes us from other animals?

[83] **Jeremy**: Yes okay, so how we acquired our moral instincts has been one of the greatest mysteries in biology. The primatologist Richard Wrangham described it as **'A question that has lain unsolved at the core of biology ever since Darwin'** (review of E.O. Wilson's 2019 book *Genesis: The Deep Origin of Societies*). And Darwin himself described it as the **'one special difficulty'** with his concept of natural selection (*On The Origin of Species*, 1859, p.209 of 440). The reason for the **'difficulty'**—and this is some basic biology for you—is that genes normally can't select for unconditionally selfless, fully cooperative traits, simply because such traits tend to be self-eliminating and so normally can't become established in a species—I mean, 'By all means, you can be selfless and sacrifice your genes for me, but I'm not about to be selfless and sacrifice

my genes for you.' Which by the way means that the 'cooperation beats competition', 'group selection' theory that left-wing thinkers say explains our capacity to be selfless is biologically impossible—see my book *Death by Dogma*. So the question is, how could such a selfish process as natural selection have created loving selflessness in us?

[84] The answer is it was achieved in our forebears through <u>nurturing</u>. To explain what is so significant about a mother's nurturing of her offspring, I first need to point out that a mother's maternal instinct to care for her offspring *is* selfish because she is ensuring the reproduction of her genes by ensuring the survival of offspring who carry her genes. So maternalism is a selfish trait, which, as I've just said, genetic traits normally have to be for them to reproduce and carry on into the next generation. HOWEVER, and this is all-important, from the infant's perspective maternalism does have the appearance of being selfless. From the infant's perspective, it is being treated unconditionally selflessly—the mother is giving her offspring food, warmth, shelter, support and protection for apparently nothing in return. So it follows that if the infant can remain in infancy for an extended period and be treated with a lot of seemingly altruistic love, it will be indoctrinated with that selfless love and grow up to behave accordingly. So selfish maternalism *can* train an infant in altruistic selflessness. (Freedom Essay 21 on our World Transformation Movement's website, and chapter 5 of *FREEDOM* explain this 'love-indoctrination' process, as we call it, more fully.)

[85] **Craig**: So what you're saying is that mothers' nurturing of their infants is primarily genetically selfish because it ensures the reproduction of her genes, but to the infant it seems like it's being given unconditionally selfless love.

[86] **Jeremy**: Yes that's right, and if we think about primates, being semi-upright from living in trees, swinging from branch to branch, and thus having their arms free to hold a dependent infant, it's clear that they are especially facilitated to support and prolong the mother-infant relationship, and so develop this nurtured, loving, cooperative behaviour.

Bonobo mothers holding their infants

[87] And in fact, bonobos, the ape species who live south of the Congo River in Africa, are extraordinarily matriarchal, or female role focused, and extraordinarily nurturing. You can find photos online—and I'll include some in the transcript booklet and video of this interview (see next page)—that illustrate just how nurturing bonobos are; they show bonobo mothers giving their infant their devoted and undivided attention!

[88] And as a result of all this nurturing, bonobos are the most co-operative and loving of all primates, which is evidenced by these absolutely amazing quotes that I just have to read to you.

[89] Bonobo zoo keeper Barbara Bell writes that **'Adult bonobos demonstrate tremendous compassion for each other…For example, Kitty, the eldest female, is completely blind and hard of hearing. Sometimes she gets lost and confused. They'll just pick her up and take her to where she needs to go'** ('The Bonobo: "Newest" apes are teaching us about ourselves', *Chicago Tribune*, 11 Jun. 1998).

[90] Primatologist Sue Savage-Rumbaugh says, **'Bonobo life is centered around the offspring. Unlike what happens among chimpanzees, all members of the bonobo social group help with infant care and share food with infants. If you are a bonobo infant, you can do no wrong…Bonobo females and their infants form the core of the group'** (Sue Savage-Rumbaugh & Roger Lewin, *Kanzi: The Ape at the Brink of the Human Mind*, 1994, p.108 of 299).

[91] A filmmaker of the French documentary *Bonobos* says, **'They're surely the most fascinating animals on the planet. They're the closest animals to man** [in that they share almost 99 percent of our genetic make-up]**…Once I got hit on the head with a branch that had a bonobo on it. I sat down and the bonobo noticed I was in a difficult situation and came and took me by the**

hand and moved my hair back, like they do. So they live on compassion, and that's really interesting to experience' (short accompanying film to the 2011 French documentary *Bonobos*).

Bonobos nurturing their infants

[92] And bonobo researcher Vanessa Woods gives this first-hand account of bonobos' unlimited capacity for love from her study of them in their home in the Congo basin: **'Bonobo love is like a laser beam. They stop. They stare at you as though they have been waiting their whole lives for you to walk into their jungle. And then <u>they love you with such helpless abandon that you love them back</u>. You have to love them back'** ('A moment that changed me – my husband fell in love with a bonobo', *The Guardian*, 1 Oct. 2015).

[93] **Craig**: Wow Jeremy, I mean these are *astonishing* quotes, they really are!

[94] **Jeremy**: Yes, they truly are amazing quotes—and bonobos are our closest living relatives, as mentioned they share 99% of our DNA. So we can see that bonobos provide the perfect evidence for how our distant ape ancestors became cooperative and loving.

[95] I have another picture here of a group of bonobos resting in a grassy glade, which I will also include in the transcript booklet, and it perfectly equates with the description I mentioned earlier that Plato gave about what life was like for humans back in the 'Golden Age' of nurtured togetherness. Plato said, **'And they dwelt naked, and mostly in the open air, for the temperature of their seasons was mild; and they had no beds, but lay on soft couches of grass, which grew plentifully out of the earth.'** Clearly we have a perfect instinctive memory (if we don't choose to deny it) of what life was like before 'the fall' because Plato didn't know of the existence of bonobos and yet knew exactly what our bonobo-like life before 'the fall' was like.

Bonobos resting in a grassy glade

[96] Now this quote is a bit long but it's such a wonderful intuitive remembrance of our species' bonobo-like time in an alienation-free, all-sensitive and all-loving state of innocence that I just have to read it out. It's from the great Russian novelist Fyodor Dostoevsky. He wrote of a time when: **'The grass glowed with bright and fragrant flowers. Birds were flying in flocks in the air, and perched fearlessly on my shoulders and arms and joyfully struck me with their darling, fluttering wings. And at last I saw and knew the people of this happy land. They came to me of themselves, surrounded me, kissed me. The children of the sun, the children of their sun — oh, how beautiful they were!...Their faces were radiant...in their words and voices there was a note of childlike joy...It was the earth untarnished by <u>the Fall</u>; on it lived people who had not sinned...They desired nothing and were at peace; <u>they did not aspire to knowledge of life as we aspire to understand it</u>, because their lives were full. But their knowledge was higher and deeper than ours...but I could not understand their knowledge. They showed me their trees, and I could not understand the intense love with which they looked at them; it was as though they were talking with creatures like themselves...and I am convinced that the trees understood them. They looked at all nature like that — at the animals who lived in peace with them and did not attack them, but loved them, <u>conquered by their love</u>...There was no quarrelling, no jealousy among them...for they all made up one family'** (*The Dream of a Ridiculous Man*, 1877).

Fyodor Dostoevsky (1821–1881)

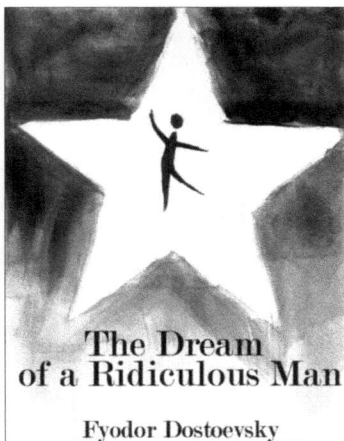

The Dream of a Ridiculous Man

Fyodor Dostoevsky

[97] This description of being **'conquered by their love'** is so like the description just given by the bonobo researcher Vanessa Woods, when she said bonobos **'love you with such helpless abandon that you love them back. You have to love them back'**. Again we see how accurate our memory is, if we don't deny it, of what life was like before **'the Fall'**.

[98] **Craig**: That quote is *so* wonderful, it's just amazing, thank you Jeremy, I'm glad you read it out. This nurturing explanation for our moral instincts seems reasonably obvious, and evidenced by those quotes about bonobo behaviour and also by the photographs of the bonobos, so why haven't I heard about this until now?

[99] **Jeremy**: Well, just like the obvious truth that our species once lived cooperatively and lovingly, this truth that we acquired our moral instincts through nurturing has been an unbearable truth while we couldn't explain why we humans became angry, egocentric and alienated and as a result lost the ability to adequately nurture our offspring with unconditional selflessness or love. The truth of our species' Edenic all-loving and all-sensitive innocent past, and the truth that nurturing is what made us human, have both been impossible truths to accept while we couldn't truthfully explain our present immensely corrupted human condition, explain why our species became so corrupted and lost the ability to fully nurture its offspring. As it's been observed, **'parents would rather admit to being an axe murderer than a bad mother or father'**! (John Marsden, *Sunday Life*, *The Sun-Herald*, 7 Jul. 2002).

[100] In fact, this reasonably obvious nurturing explanation for our moral conscience was first put forward by the American philosopher John Fiske in his book *Outlines of Cosmic Philosophy*, which was published in 1874, only a few years after Darwin published his theory of natural selection. And, at the time, Fiske's explanation was actually recognised as being, and I quote, **'far more important'** than **'Darwin's principle of natural selection'** and **'one of the most beautiful contributions ever made to the Evolution of Man'** (Dorothy Ross, *G. Stanley Hall: The Psychologist as Prophet*, 1972, p.262 of 482). And Darwin himself went so far as to write to Fiske saying, **'I never in my life read so lucid an expositor (and therefore thinker) as you are'** (1874; *Life and Letters of Charles Darwin, Vol. 2*). But again,

while we couldn't explain our loss of ability to adequately nurture our offspring, this **'far more important'** insight than **'Darwin's principle of natural selection'** was let die and eventually disappeared from biological discourse!

OUTLINES OF COSMIC PHILOSOPHY

John Fiske (1842–1901)

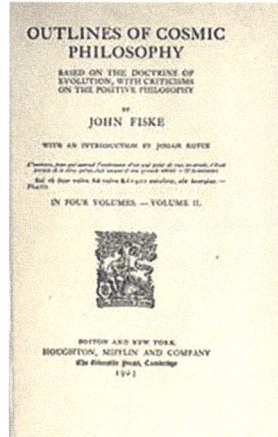

[101] I might point out that 1) Darwin's natural selection explanation for the variety of species; 2) Fiske's and my nurturing explanation for our moral instincts; 3) the instinct vs intellect explanation that I've given for the human condition; 4) the explanation I give in chapter 7 of *FREEDOM* for how we humans became fully conscious when other species haven't; and 5) the Negative-Entropy-driven integrative meaning of existence (which we have personified as 'God') that I explain in chapter 4 of *FREEDOM*—which together are the five main questions science has had to answer about our world and place in it—are all reasonably obvious, straightforward, simple explanations that bear out biologist Allan Savory's observation that **'whenever there has been a major insoluble problem for mankind, the answer, when finally found, has always been very simple'** (*Holistic Resource Management*, 1988, 1st edition, p.3). As *FREEDOM* makes clear, having to live in denial of the human condition has blocked our access to so many reasonably obvious truths.

[102] **Craig**: Thanks Jeremy. This is really amazing to learn about how science has denied our species' cooperative and loving heritage, and the nurtured origins of that cooperative and loving soul. But it all really does make sense—I mean <u>we certainly needed the defence for our corrupted condition before we could face the truth about it</u>.

[103] I am speaking with Australian biologist Jeremy Griffith.

Part 4

How does the psychological rehabilitation of the human race that the arrival of understanding of the human condition finally makes possible actually occur?

[104] **Craig**: Hello and welcome back to Part 4 of the interview. I'm Craig Conway and I'm here with Jeremy Griffith, the biologist who has just explained how humans acquired our moral instincts. This is an absolutely remarkable interview!

[105] So continuing on, Jeremy, what about my third question for this final Part 4 of the interview, which is how does **'the psychological rehabilitation of the human race'** that this understanding gives us actually take place; do we all need to go into therapy or something?

[106] **Jeremy**: Well, what this real—and actually very obvious—instinct vs intellect explanation of the human condition fundamentally does is lift the 'burden of guilt' from the human race. It establishes that we humans are good and not bad after all. While we are all inevitably variously angry, egocentric and alienated from our different encounters with humanity's heroic battle to find knowledge, ultimately self-knowledge, understanding of our corrupted condition, <u>we can now know that every human is fundamentally good</u>. And this ability to understand and know there was a good reason why the human race became psychologically upset, is the key, relieving understanding we have been in search of ever since we became conscious some 2 million years ago and our corrupted condition emerged.

[107] That is the key relief for our mind—being finally able to understand that we are good and not bad is what brings us the greatest psychological relief of all. The psychoanalyst Carl Jung said, **'wholeness for humans depends on the ability to own our own shadow'**, and since we can now **'own'** the **'shadow'** of our species' 2-million-year-corrupted condition, the human race *is* finally in a position to become **'whole'**. The word '<u>psychosis</u>' literally means **'soul-illness'** and '<u>psychiatry</u>'

literally means **'soul-healing'** (derived as they are from *psyche* meaning 'soul', *osis* meaning **'abnormal state or condition'** and *iatreia* meaning 'healing'—see pars 63 & 72 of *FREEDOM*), but we have never been able to 'heal our soul', explain to our original instinctive self or soul that we, our fully conscious thinking self, is good and not bad and by so doing reconcile and heal our split selves—but now at last we can.

[108] **Craig**: Well, there is an adage that says **'The truth will set you free'**, so what you're saying then Jeremy is that the truth of our fundamental goodness is the truth that we needed to set us free from the human condition.

[109] **Jeremy**: Precisely, and while that is the main relief our mind needed, obviously the more we digest that relieving understanding, the more healing relief comes to *every* aspect of our upset condition—and to have had to endure being unjustly condemned as bad for 2 million years does mean there is a great deal of upset to heal.

[110] To appreciate how much upset exists in us humans now, imagine living for just one day with the injustice of being condemned as bad, even evil, when you intuitively knew but were unable to explain that you were actually the complete opposite of evil, namely truly wonderful, good and meaningful—in fact not just good but the *hero of the story of life on Earth*! You would be hurt to the core and furious wouldn't you! Now extrapolate that experience over 2 million years and we can begin to appreciate just how much volcanic frustration and anger must now exist within us humans! While we have learnt to significantly restrain and conceal—'civilise' as we refer to it—the phenomenal amount of upset within us, under the surface we all must be boiling with rage, and sometimes, when our restraint can no longer find a way to contain it, that anger must express itself—hence our capacity for shocking acts of cruelty, sadism, hate, murder and war.

We humans aren't bad but we are certainly mad!

[111] And no wonder we have led such an evasive, denial-practising, lying, avoid-any-criticism, escapist, <u>alienated</u>, superficial and artificial, greedy, <u>egocentric</u>, power, fame, fortune and glory-seeking existence. We have *had* to smother ourselves with *material* glory while we lacked the *spiritual* glory of compassionate understanding of ourselves.

Material reinforcement had to sustain us until we found spiritual reinforcement, understanding

[112] So there is an enormous amount of upset to subside and heal in us humans, and that will obviously take time. In fact, we have to expect that it will take a number of generations to be completely ameliorated. But the good news, and this is very important, is that while it will take a number of generations to heal all the upset in us humans, <u>everyone can immediately live free of their upset</u>. The reason we can live free of it is that while we lacked the real defence and reinforcement of understanding of our corrupted condition, we absolutely needed the artificial defences and reinforcements of attacking any criticism of our corrupted condition, of denying and blocking it out, and of finding any positive reinforcement we could—anger, alienation and egocentricity are what sustained us—but now that we have the real defence and re-inforcement of our fundamental goodness, all these artificial defences and reinforcements are obsoleted, they are no longer needed. <u>In fact, to continue using the old artificial defences of retaliation, denial and the search for relieving power, fame, fortune and glory when our funda-mental goodness has been established is not only clearly pointless but also unnecessarily destructive of ourselves, everyone around us and our planet.</u> That way of living is now completely obsoleted, finished with.

[113] **Craig**: That makes total sense; I mean our artificial ways of reinforcing ourselves are obsoleted by the real reinforcement of our-selves. One way of living ends, a new one begins, free of the human condition—phew, thank God for that!!

[114]**Jeremy**: That's for sure! Basically, now that our corrupted condition is finally truthfully explained, honesty replaces denial and the world heals. 'Bullshit', which is our everyday word for all the dishonest denial that's been going on, protected us from all the truths about our corrupted condition that we couldn't properly explain, but it was destroying the world and it now stops.

[115] With this end of lying in mind, there's one more thing I should explain, which is that socialism, the new age movement, the politically correct 'woke' movement, and all the other idealistic movements, were actually all <u>false starts</u> to a human-condition-free world because the upsetting battle to find knowledge, ultimately self-knowledge, the psychologically relieving understanding of why we are good and not bad, still had to be completed. In fact, while dogmatically insisting that everyone should be cooperative and loving could make you feel that you were doing good and be superficially psychologically very relieving, such insistence on ideal behaviour denied people the freedom of expression and individualism they needed to be able to continue the all-important upsetting search for knowledge. These movements were <u>pseudo idealistic movements</u> that stifled and oppressed the search for the understanding of our corrupted condition that was needed to *actually* free us from that state. They were regressive, not progressive as they deluded themselves they were. It was actually the right-wing who have supported the upsetting battle to find knowledge that held the moral highground, not the pseudo idealistic left-wing.

[116] The culture of the Left made people superficially feel good but it was dangerously dishonest, it was fake—it was bullshit. Being concerned for others and the world is *very* important, but doing that to make yourself feel good is a dangerously selfish sickness, indeed it's the most destructive of all drug addictions—and it's been taking over the world. As I explain in my free book, *Death by Dogma*, dogma is not the cure, it's the poison.

[117] You can see here that the true instinct vs intellect explanation of the human condition finally enables us to explain and expose what's wrong with the Left—and it's not a moment too soon because its culture is rapidly taking our species to death-by-dogma extinction.

We 2-million-year psychologically upset humans needed to be able to think our way to sanity. Suppressing freedom of thought turned humans into brainless robots destined to march off the edge of hope into the chasm of terminal psychosis and extinction.

[118] Now, most wonderfully of all, the instinct vs intellect explanation of the human condition not only exposes the culture of the Left for the human-race-destroying lie that it is, it also, as I said, brings to an end the whole upsetting search for the rehabilitating understanding of why we're good and not bad, and what this means is that it's no longer oppressive of that upsetting search to take up support of cooperative and loving idealism—because that search is over. In fact, taking up support of cooperative and loving idealism is now the only way to live that's justified! Suddenly there's no longer any reason for the right-wing in politics and everyone effectively becomes left-wing. In fact, the whole business of politics basically ends with the finding of understanding of the human condition, and the whole human race sets out as one united organism letting go the angry, egocentric and aliena-ted part of ourselves and supporting cooperative, selfless and loving idealism. (Much more is explained about pseudo idealism in *Death by Dogma* on the World Transformation Movement's website.)

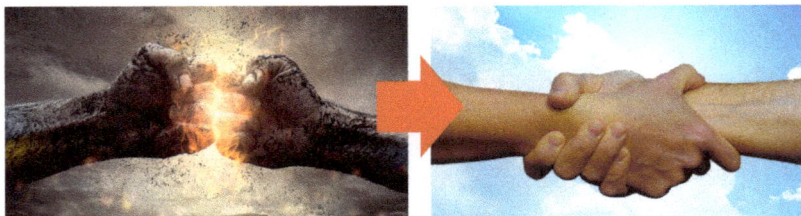

[119] **Craig**: Yes, ending the polarised world of politics will certainly be one of the biggest reliefs imaginable!

[120] **Jeremy**: Absolutely Craig. It will be a massive relief.

[121] So that's how the whole world suddenly, immediately changes from a psychologically embattled angry, egocentric and alienated state, to a world where everyone has decided to abandon their still-to-be-healed competitive and aggressive behaviour and takes up support of a cooperative and loving existence. [See www.humancondition.com/transformation for a full description of this great transformation.]

[122] So finding understanding of the human condition brings to an end the insecure, upset, artificial-reinforcement-dependent angry, egocentric and alienated world. A new human-condition-resolved, cooperative, selfless and loving world now emerges. Light comes streaming into the dark cave-like world of denial that we have been living in, and it will all be like waking up from a nightmare!

[123] Basically, with the ability to understand ourselves, we can return to our original cooperative and loving state, but this time fully conscious. As the poet T.S. Eliot anticipated, **'We shall not cease from exploration and the end of all our exploring will be to arrive where we started and know the place for the first time'** (*Little Gidding*, 1942).

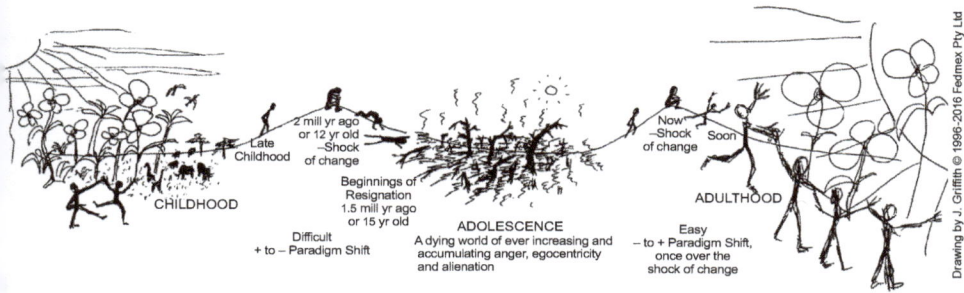

Humanity's Journey from Ignorance to Enlightenment, see chapter 8 of *FREEDOM*.

Australopithecus afarensis Fossil evidence from 3.9 to 3 million years ago Brain Volume 400 cc average	*Australopithecus africanus* 3.3 to 2.1 m y a 450 cc	*Australopithecus boisei* 2.3 to 1.2 m y a 530 cc	Human Condition Fully Emerges Here	*Homo habilis* 2.4 to 1.4 m y a 650 cc	*Homo erectus* 1.9 to 0.1 m y a 900-1100 cc	*Homo sapiens* 0.5 to 0.1 m y a 1350 cc	*Homo sapiens sapiens* 0.2 m y a to now 1400 cc
Early Happy Childman	Middle Demonstrative Childman	Late Naughty Childman		Distressed Adolescentman	Adventurous Adolescentman	Angry Adolescentman	Pseudo idealistic and Hollow Adolescentman

Humanity's stages of maturation, see chapter 8 of *FREEDOM*.
(Note, our large brain appeared some 2 mya – see earlier reference in par. 63.)

[124] **Craig**: Wow Jeremy, that has been absolutely incredible, enlightening, enthralling, and I really can't thank you enough for sharing with us your knowledge and your insight. To think about the human race

being transformed just in the nick of time is, I think, what everybody out there who has listened to this will be hoping for.

[125] And your book *FREEDOM*, again, is available on HumanCondition. com for everybody to access—and it's free on there as well. So please follow this interview up, listen to it again and again, try your best to understand what Jeremy is saying here, but get online, get the information and keep studying, and let's hope we can all enjoy and embrace a new change for the world, for all of us.

[126] Jeremy, it's been wonderful to talk with you. Thank you for joining us, and we will hopefully speak to you again—but even better, that we'll be seeing your work in the lives of everybody across the planet very soon.

[127] I'm Craig Conway, this has been my interview with Jeremy Griffith.

END OF INTERVIEW

Genevieve Salter: So everyone, don't forget you can learn all about this fabulous, world-saving breakthrough understanding of the human condition at HumanCondition.com—including a description of how the psychological rehabilitation of humans actually occurs, and how everyone's life can immediately be transformed at www. humancondition.com/transformation!!

And I want to mention that what you'll see there following this interview is a series of videos that elaborate on what's been outlined in the interview—so there'll be some repetition, but going over the concepts in more detail does help with the understanding and absorption of them.

In the first of those videos Jeremy warns us of our historic fear of the subject of the human condition and the resulting 'deaf effect' that reading or hearing about the human condition often initially causes.

**Visit www.HumanCondition.com to view
the film of this interview, and to watch the series
of introductory videos Genevieve has referred to.**

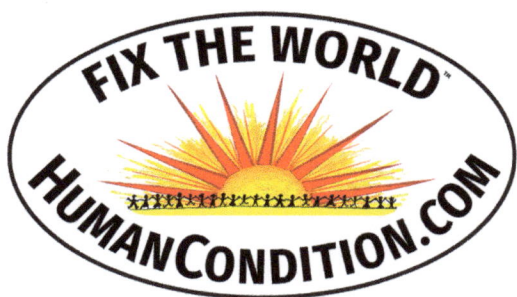

FIX THE WORLD™
HumanCondition.com

With the real problem of the human condition finally solved we can now ACTUALLY fix the world!